JOURNEY THROUGH THE MESOZOIC ERA

Copyright © 2012 por Flora Goldberg.

ISBN: Tapa Blanda 978-1-4633-3408-6
Libro Electrónico 978-1-4633-3409-3

Todos los derechos reservados. Ninguna parte de este libro puede ser reproducida o transmitida de cualquier forma o por cualquier medio, electrónico o mecánico, incluyendo fotocopia, grabación, o por cualquier sistema de almacenamiento y recuperación, sin permiso escrito del propietario del copyright.

Este Libro fue impreso en los Estados Unidos de América.

Para hacer pedidos de copias adicionales de este libro, por favor contactar con:
Palibrio
1663 Liberty Drive, Suite 200
Bloomington, IN 47403
Para llamadas desde los EE.UU 877.407.45847
Para llamadas internacionales +1.812.671.9757
Fax: +1.812.355.1576
ventas@palibrio.com
[416592]

JOURNEY THROUGH
THE MESOZOIC ERA

ILLUSTRATIONS
FLORA GOLDBERG

A JOURNEY THROUGH THE MESOZOIC ERA

Mesozoic means middle life.

The mesozoic era is an interval of geological time, from about 250 million years ago to 65 million years ago, it is also call the age of the dinosaurs because of it´s association with non-avian dinosaurs.

During the mesozoic era the earth was very different than it is now, the climate was warmer, the seasons were very mild, the sea level was higher and there were not polar ice.

Even the shape of the continents was different, they jammed together at the beginning of the mesozoic forming the Supercontinent of Pangaea.

The Mesozoic era was a time of tectonic, climatic and evolutionary activity. The gradual drift of the continents toward their present position resulted in the end of the Supercontinental Pangaea

The Chicxulub impact and other events ended the era when a majority of species on earth went extint.

At the end of the era the basis of the modern life was started.

The mesozoic Era is divided in 3 big periods:

TRIASSIC Period 248- 206 M.Y.A.

JURASSIC PERIOID 206-144 M.Y.A.

CRETACEOUS PERIOD 144-65 M.Y.A.

JOURNEY THOROUGH THE MESOZOIC ERA

TRIASSIC PERIOD 248-206

Is named Triassic Period for the kind of rocks that were depositated during this period.
It is the first period of the Mesozoic Era, it is in many ways a time of transition, it was a this time that the World Continent of
Pangea existed altering global climate and ocean circulation. The Triassic also follows the largest extinction event in the history of life, and so it is a time when survivors of that event spread and recolonized.
It is a fascinating period because of the animals that lived in the planet "The Dinosaurs".
At the beginning of this period the wether was hot and dry and they were big desertic areas, ideal for the reptils and carnivores bipeds, and the plants were Cactaceas and Coniphers.
But it was during the late Triassic when Pangaea start to break in 3 parts: Eurasia with North América, Zafria with South América and Antártida with Australia and India.
with the appearence of the Pacific Ocean an important change of weather occure, and with higher temperatures and humidity it became an ideal scenery for the proliferation of Dinosaurs.

Subdivision of the Triassic:
Inferior: Olenekiense, induense
Medium::   Ladinense, Luisiense
Superior: Platinense, Noriense

Walkeria
Early Triassic 231 M.Y.A
Carnivore
India

flora goldberg

Tanystropheus
Middle Triassic 232 M.Y.A.
Carnivore 20 feet long
Europe, Middle east.

flora goldberg

JOURNEY THOROUGH THE MESOZOIC ERA

Proterosuchus
Early Triassic 300 M.Y.A.
carnivore reptile
South Africa, China

flora goldberg

Riojasaurus
Early Triassic 299 M.Y.A.
Herbivore 30 feet long
Madagascar, Marrocos, Argentina

flora goldberg

JOURNEY THROUGH THE MESOZOIC ERA

Edaphosaurus
Early Triassic 303 M.Y.A.
Herbivore
N.A., Slovakia

flou goldberg

Barapasaurus
Late Triassic 189-176 M.Y.A
Herbivore 50 feet long
India

flou goldberg

Euskelosaurus
Late Triassic 217 M.Y.A
Herbivore 30 feet long
South Africa, Zimbabwe

flora goldberg

Nothosaurus
Late Triassic 240-210 M.Y.A.
Carnivore.
from N.A, Europe, China

flora goldberg

JOURNEY THROUGH THE MESOZOIC ERA

Eoraptor
Late Triassic 228 M.Y.A
Carnivore 3 feet long
Argentine

flora goldberg

Lilienstervus
Late Triassic 205 M.Y.A
Carnivore
Argentina

flora goldberg

JOURNEY THOROUGH THE MESOZOIC ERA

Peteinosaurus
Late triassic 215.M.Y.A.
carnivore
Italian alpes

flora goldberg

Pisanosaurus
Late Triassic 231 M.Y.A
Herbivore 4 feet long
Argentine

flora goldberg

JOURNEY THROUGH THE MESOZOIC ERA

Mussaurus
Late Triassic - 227 M.Y.A
Herbivore 9 feet long
Argentina

flora goldberg

Pantydraco
Late Triassic
Omnivore
England.

flora goldberg

10

JOURNEY THOROUGH THE MESOZOIC ERA

Proganochelys
Late Triassic 210 M.Y.A
Omnivore 3 ft. long
Germany Thailand

flora goldberg

Staurikosaurus
Late Triassic 220 M.Y.A
Carnivore 7 4 feet long
Brazil

flora goldberg

JOURNEY THROUGH THE MESOZOIC ERA

Procompsognathus
Late Triassic 222 M.Y.A.
Carnivore 4 feet long
Germany

flora goldberg

Plateosaurus
Late Triassic 199 M.Y.A.
Herbivore 18 feet long
Germany

flora goldberg

Azendohsaurus
Late Triassic 227 M.Y.A
Omnivore 6ft. Long
Morroco Marakesh

flora goldberg

Coelophysis
Late triassic 216-205 M.Y.A.
biped carnivore U.S.A.

flora goldberg

JOURNEY THROUGH THE MESOZOIC ERA

Zatomus
211th Late Triassic
Carnivore
North Carolina

flora goldberg

Thecodontosaurus
Late Triassic 216-200 M.Y.A
Herbivore
South England

flora goldberg

JOURNEY THOROUGH THE MESOZOIC ERA

Sellosaurus
Late Triassic 200 M.Y.A.
Herbivore 10 feet long
Germany

flora goldberg

Revueltosaurus
Late Triassic 225 M.Y.A
Herbivore
New Mexico U.S.A.

flora goldberg

JOURNEY THROUGH THE MESOZOIC ERA

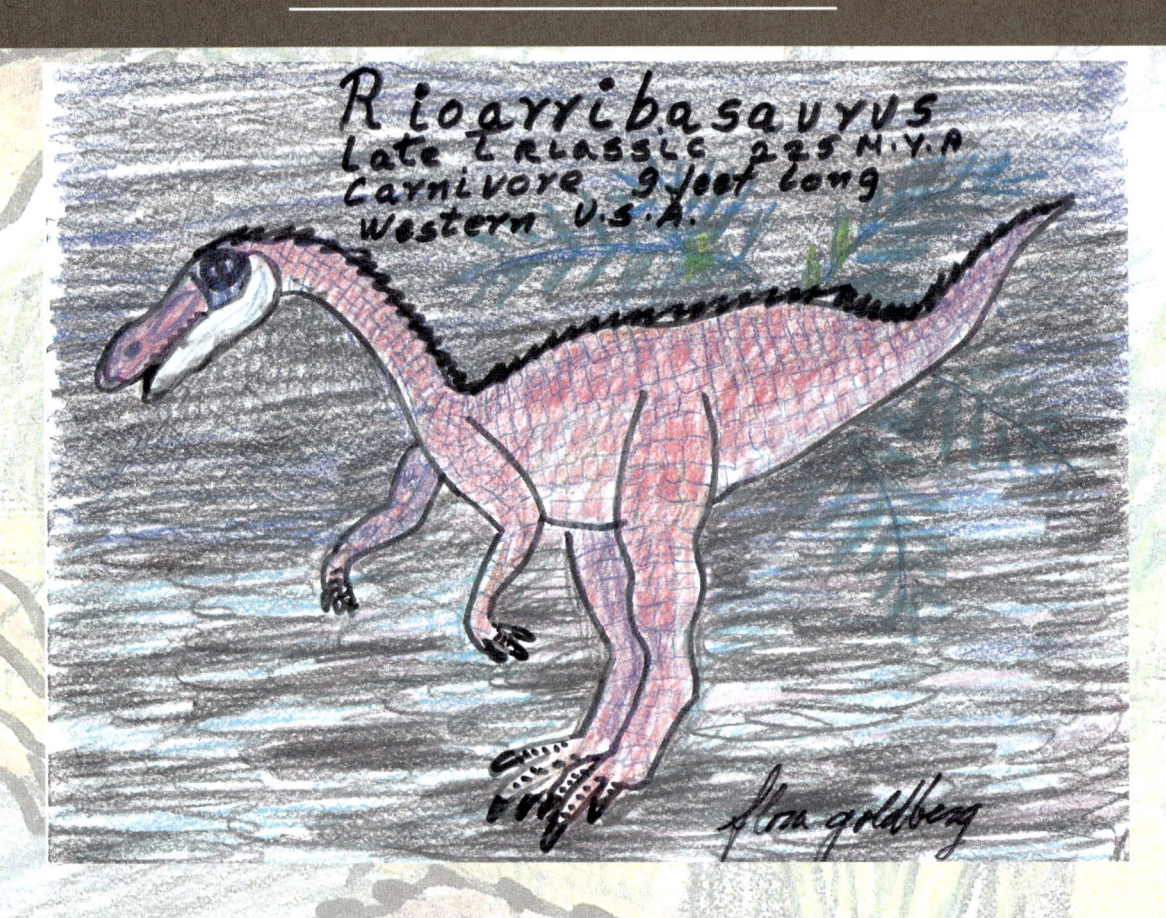

Rioarribasauyus
Late Triassic 225 M.Y.A
Carnivore, 9 feet long
Western U.S.A.

flora goldberg

Staurikosaurus
Triassic late 225 M.Y.A.   Carnivore 7ft long
S.A.

flora goldberg

16

Jurassic Period 206-144 M.Y.A.

Named for the Jura Mountains on the border between France and switzwerland, where rocks of this age were studied, is also know as the age of the reptils.

By the beginning of the jurassic period the Supercontinental Pangaea had begin rifting into two landmasses; Laurasia to the North, and Gondwana to the South, that created more coastlines and shifted the continental climate from dry to humid, and many of the desserts of the triassic were replaced by lush rainforest. Great plant eating Dinosaurs roaming the earth, feeding on lush growths of ferns and palm-like cycads and bennettilaleans, smaller but vicious carnivores stalking the great herbivores, oceans full of fish, squid and coiled ammonites, plus great Ichthosaurus and long necked Plesiosaurus, vertebrates taking the air, like the ptesosaurus and the first birds, that was the Jurassic period...

Subdivision of the Jurassic.
Early: Hettongean, Sinenurian, Pleibashian, Toorcian
Middle: Aalenian, ,Bajocian, Ballhonian, Callovian.
Late;   Oxfordian, Kimmeridgian, Thithonian

JOURNEY THROUGH THE MESOZOIC ERA

Dilophosaurus
Early Jurassic 190 M.Y.A.
Carnivore 20ft. long
N.A.   Asia

flora goldberg

Heterodontosaurus
Early Jurassic 195 M.Y.A
Herbivore 5ft. long
Africa

flora goldberg

JOURNEY THOROUGH THE MESOZOIC ERA

Fabrosaurus
Early Jurassic 208 M.Y.A
Herbivore 33% long
South Africa

flora goldberg

Massospondylus
Early Jurassic 203-183 M.Y.A.
Herbivore 13-20 feet long
Africa, Zimbabwe, U.S.A.

flora goldberg

JOURNEY THROUGH THE MESOZOIC ERA

Lesothosaurus
Early Jurassic 199 M.Y.A
Herbivore 3 f. long
South Africa

Yunnanosaurus
Early Jurassic 190 M.Y.A.
Herbivore 21 feet Long
Yunnan, China

JOURNEY THOROUGH THE MESOZOIC ERA

Segisaurus
Early Jurassic 190 M.Y.A.
Carnivore 4 feet long
N.A.

flora goldberg

Anchisaurus
Early Jurassic
Herbivore 6.6 fee long
N.A.

flora goldberg

JOURNEY THROUGH THE MESOZOIC ERA

Stenopterygus
Early Jurassic 183 M.Y.A.
Carnivore   12 feet long
England, France, Germany

flora goldberg

Scelidosaurus
Early Jurassic 208-194 M.Y.A.
Herbivore   13 feet long
Europe, N.A.

flora goldberg

JOURNEY THOROUGH THE MESOZOIC ERA

Zigongosauyus
Middle Jurassic 168 M.Y.A.
Herbivore 52 f. long
Sechuan China

flora goldberg

Ornitholestes
Late Jurassic 145 M.Y.A.   Carnivore 6½ f. long
U.S.A.

flora goldberg

JOURNEY THROUGH THE MESOZOIC ERA

Abyosauyus
Middle Jurassic 168 M.Y.A
Herbivore 30 ft. long
CHina

Coeluyus
Late Jurassic 150 M.Y.A
Carnivore 6 ft. long
Wyoming U.S.A.

JOURNEY THOROUGH THE MESOZOIC ERA

Diplodocus
Late Jurassic 150 M.Y.A
Herbivore 111 ft long 20ft high
N.A.

flora goldberg

Syntarsus Kayentatas
Early Jurassic   206 M.Y.A
Carnivore   10 feet long
Africa Zimbabwe, U.S.A.

flora goldberg

25

JOURNEY THROUGH THE MESOZOIC ERA

Yangchuanosaurus
Late Jurassic 150 M.Y.A
Carnivore
Mongolia

flora goldberg

Stegosaurus
Late Jurassic 156 M.Y.A
Herbivore
Portugal west U.S.A.

flora goldberg

26

JOURNEY THOROUGH THE MESOZOIC ERA

Camptosaurus
Late Jurassic 152 M.Y.A.
Herbivore 21 ft. long
N.A Europe

flora goldberg

Kentrosaurus
Late Jurassic 152 M.Y.A.
Herbivore 12 feet long 320 kilos
Africa.

flora goldberg

JOURNEY THROUGH THE MESOZOIC ERA

Jenghizkhanosaurus
155 M.Y.A Late Jurassic
Carnivore
Mongolia

flora goldberg

Ceratosaurus
Late Jurassic 152 M.Y.A
Carnivore 18 f. long
N.A Africa Europe

flora goldberg

Allosaurus
Late Jurassic 156 M.Y.A.
carnivore
N.A., Africa, Australia, China

flora goldberg

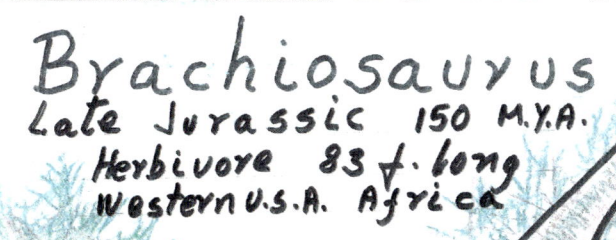

Brachiosaurus
Late Jurassic 150 M.Y.A.
Herbivore 83 f. long
Western U.S.A. Africa

flora goldberg

JOURNEY THROUGH THE MESOZOIC ERA

Archaeopteryx
Late Jurassic 155 M.Y.A
Carnivore
Germany

flora goldberg

Camarasaurus
Late Jurassic 155 M.Y.A.
Herbivore 54 feet long 18 tons.
N.A.

flora goldberg

JOURNEY THOROUGH THE MESOZOIC ERA

Plesiosauyus
Late Jurassic 135 M.Y.A.
Carnivore 11.5ft long
England, Germany

flora goldberg

Lenonlosauyus
115M. Early Cretaceous
Herbivore
U.S.A.

flora goldberg

Cretaceous Period 144-65 M.Y.A.

The Cretaceous period is usually noted for being the last period of the age of Dinosaurs.

The break-up of the world Continent Pangaea to the actual shapes, the weather became also similar to the actuals, and made possible the life of many insects groups, modern mammals and bird groups and the first flowering plants.

The break-up of the world Continent Pangaea which began to disperse during the Jurassic continued, this lead to increase regional differences in floras and faunas between the Northen and Southern continents.

The end of the Cretaceous brought the end of previously succesful groups of organism, and laid open the stage for those groups which had previously taken secondary roles.

Cretaceous was the time in which life as it is now on earth came together.

The Chicxculub impact and other events ended this era and the majority of the species were extint.. But the vegetal world was very little afected with the final crisis of the Cretaceous.

Subdivision of the Cretaceous.

Early: Berriasian, Valenginian, Hanterivian, Barremian, Aplian, Albian.

Late: Cenononian, Turonian, Santonian, Companion, Maastrichtion.

JOURNEY THOROUGH THE MESOZOIC ERA

Psittacosaurus
130.M. Early cretaceous
Herbivore
China Mongolia

Ouranosaurus
Cretácico Temprano 100 M.A.
Herbivoro

flora goldberg

JOURNEY THROUGH THE MESOZOIC ERA

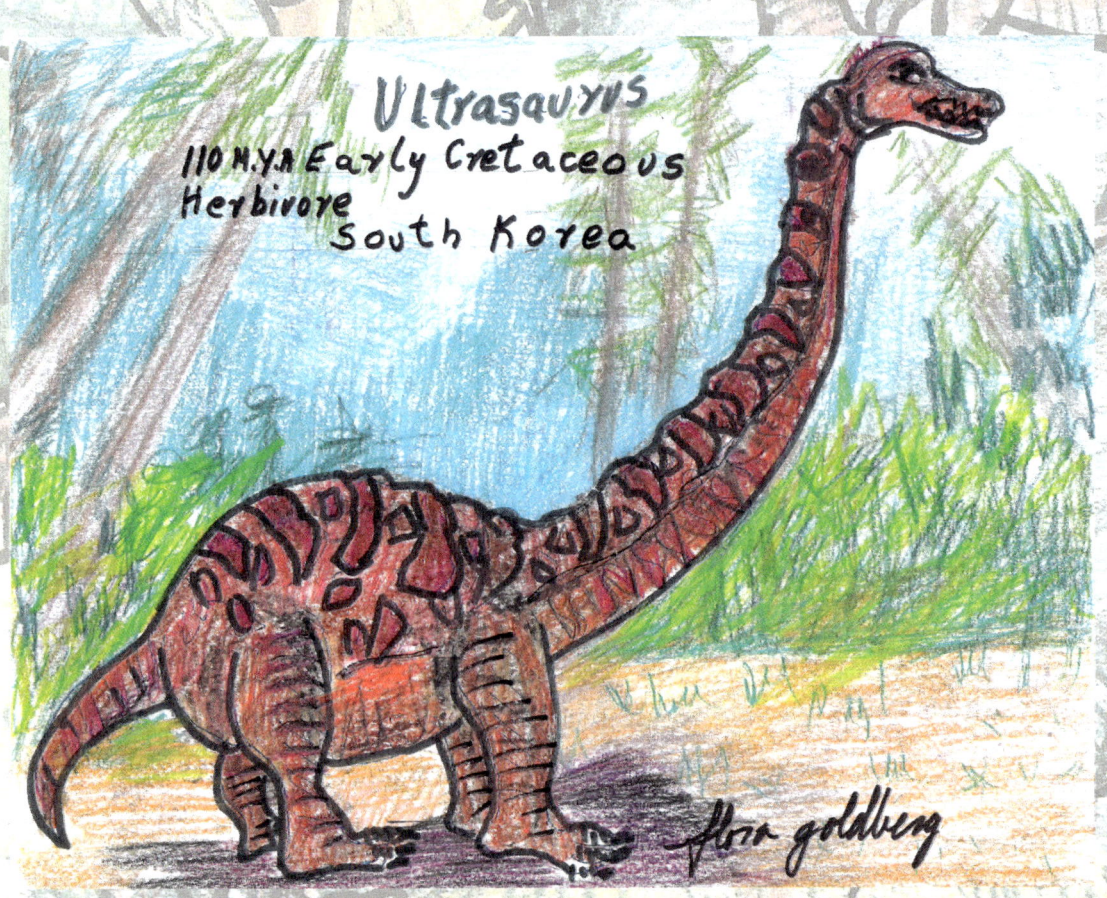

Muttaburasaurus

Early Cretaceous 100 M.Y.A
Herbivore
Australia

flora goldberg

Ultrasauyus

110 M.Y.A Early Cretaceous
Herbivore
South Korea

flora goldberg

JOURNEY THOROUGH THE MESOZOIC ERA

Iguanodon
Herbivore
Early Cretaceus Period

Baryonyx
Early Cretaceous 130 M.Y.A
Piscivorous
Spain, Portugal, England

flora goldberg

JOURNEY THROUGH THE MESOZOIC ERA

Pentaceratops
65 M.Y.A. Late cretaceous
Herbivore
New Mexico

flora goldberg

Wuerhosaurus
Early cretaceous 130 M.Y.A.
Herbivore 19 feet long
China

flora goldberg

36

Hypacrosaurus

70 M.Y.A Late cretaceous
Herbivore
Alberta Canada
Montana U.S.A.

flora goldberg

Bactrosaurus

Late cretaceous 85 M.Y.A.
Herbivore 18 feet long 6 feet high 15 tons.
Asia

flora goldberg

JOURNEY THROUGH THE MESOZOIC ERA

Adasaurus
Late Cretaceous
carnivore 6 feet long
Mongolia

flora goldberg

Xenotarsosaurus
Late Cretaceous 83-75 M.Y.A
Carnivore 1 ton.
Argentine

flora goldberg

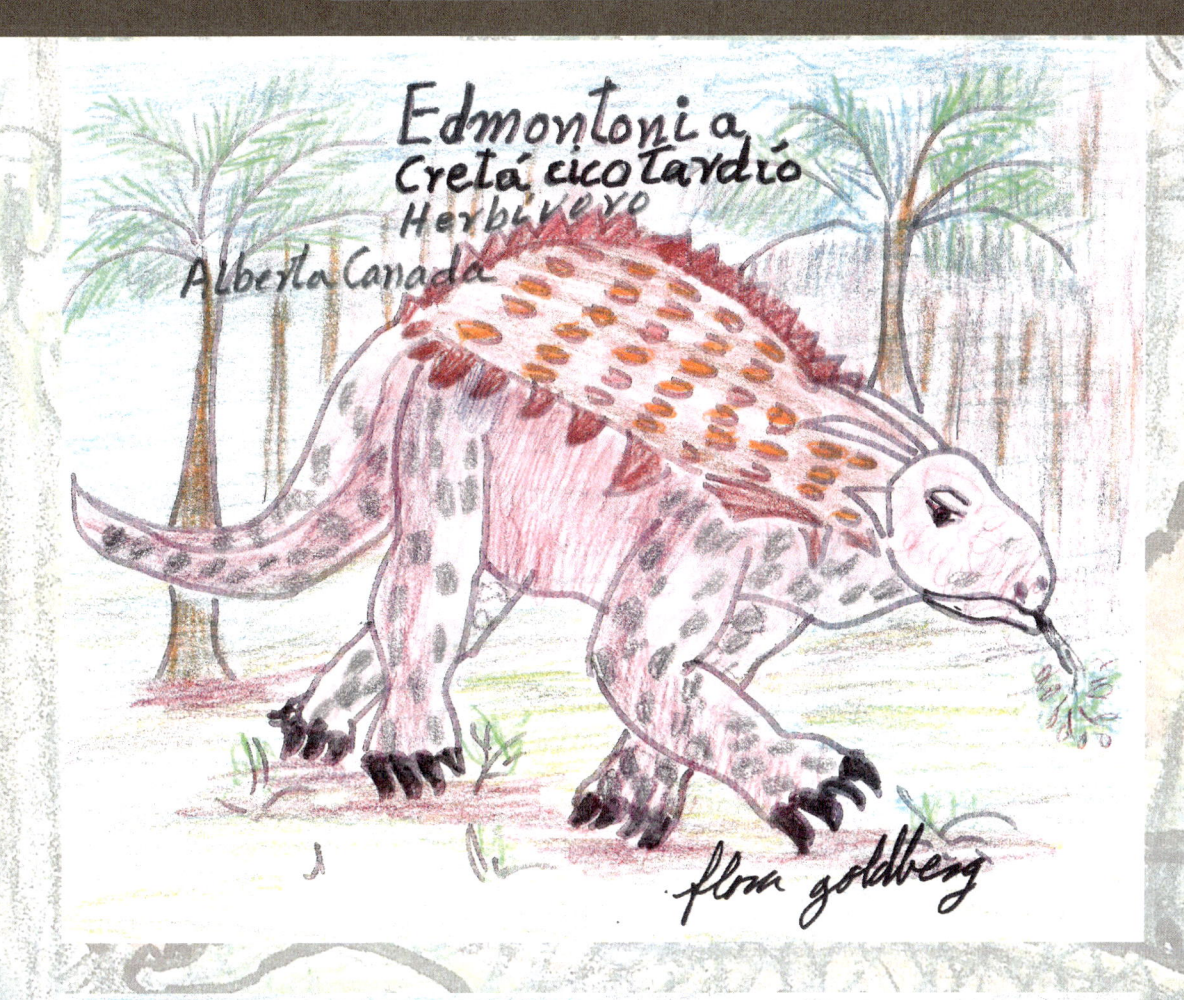

Edmontonia
Cretácico tardío
Herbívoro
Alberta Canada

flora goldberg

Tarbosaurus-boatar
Late cretaceous 70 M.Y.A
Carnivore
Mongolia China

JOURNEY THROUGH THE MESOZOIC ERA

Betasuchus
Late cretaceous 70 M.Y.A.
carnivore
Europe   Holland

flora goldberg

Anatosaurus
Late cretaceous 73 M.Y.A.
Herbivore 43 feet long
Alberta Canada

flora goldberg

Albertosaurus
Late Cretaceous 70 M.Y.A.
Carnivore 30 feet Long
Alberta Canada

flora goldberg

Gallimimus
Late Cretaceous 65 M.Y.A
Carnivore 18 feet Long
Asia Mongolia

flora goldberg

JOURNEY THROUGH THE MESOZOIC ERA

Adasaurus
Late Cretaceous
carnivore 6 feet long
Mongolia

flora goldberg

Quetzalcoatlus
Late Cretaceous   65 m.y.a.
27 feet long   fish eater
México
carnivoro
Texas, U.S.A.

flora goldberg

42

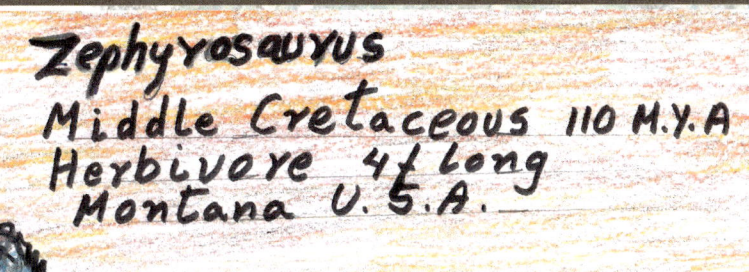

Zephyrosaurus
Middle Cretaceous 110 M.Y.A
Herbivore 4f long
Montana U.S.A.

flora goldberg

Ankylosaurus
Late Cretaceous 65 M.Y.A
Herbivore 27 feet long 6 tons.
Montana U.S.

flora goldberg

JOURNEY THROUGH THE MESOZOIC ERA

Carnotaurus
Late Cretaceous 65.M.Y.A.
Carnivore 21 feet long
Patagonia Argentine

flora goldberg

Velociraptor
Late Cretaceous 70 M.Y.A
Carnivore
Mongolia, China

flora goldberg

JOURNEY THOROUGH THE MESOZOIC ERA

Pteranodon
Late Cretaceous 73 M.Y.A
Carnivore 25 ft long
Europe

flongoldberg

Yunnanosaurus
Jurásico Temprano 180 M.A.
Herbívoro   7 mt. largo
Yunnan, CHina

flongoldberg

Corythosaurus
Late Cretaceous 77.M.Y.A
Herbivore 27 f long
North America

flora goldberg

Kronosaurus
late cretaceous 99.M.Y.A
marines carnivores
Australia

flora goldberg

JOURNEY THOROUGH THE MESOZOIC ERA

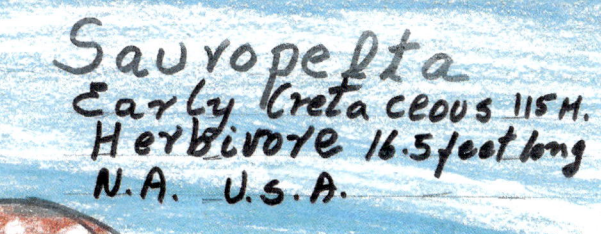

Sauropelta
Early Cretaceous 115 M.
Herbivore 16.5 feet long
N.A. U.S.A.

flora goldberg

Oviraptor
Late cretaceous 75 M.Y.A.
Omnivore
Mongolia

flora goldberg

47

JOURNEY THROUGH THE MESOZOIC ERA

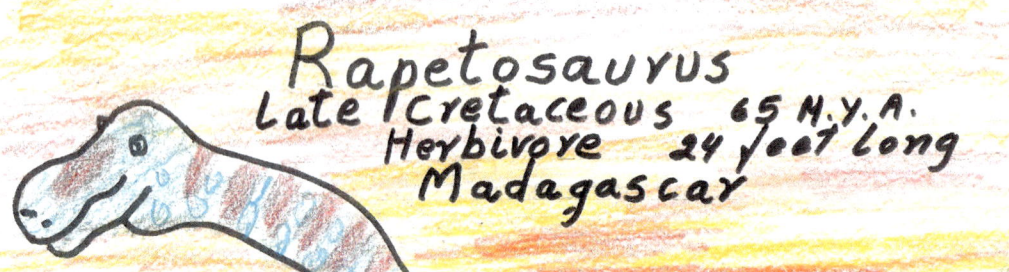

Rapetosaurus
Late Cretaceous 65 M.Y.A.
Herbivore 24 feet long
Madagascar

Majungasaurus
Late Cretaceous 65 M.Y.A.
Carnivore   Egypt
           Madagascar

JOURNEY THOROUGH THE MESOZOIC ERA

Late cretaceous

Centrosaurus
75 M.Y.A.
Herbivore
canada Alberta

flora goldberg

Craspedodon
Late cretaceous 80 M.Y.A.
Carnivore
Mongolia

flora goldberg

49

Made in the
USA
Middletown, DE